JOWETT
1901–1954

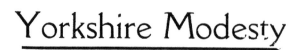

JOWETT
1901–1954

NOEL STOKOE

The
History
Press

First published in 1999
Reprinted 2000, 2002, 2010, 2022

The History Press
97 St George's Place,
Cheltenham, Gloucestershire, GL50 3QB
www.thehistorypress.co.uk

Copyright © Noel Stokoe, 2010

ISBN 978 0 7524 1723 3

Typesetting and origination by
The History Press
Printed by TJ Books Limited, Padstow, Cornwall

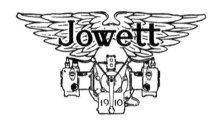

BIBLIOGRAPHY

The Complete Jowett History   Paul Clark and Edmund Nankivell
The Jowett Jupiter, The Car that Leapt to Fame   Edmund Nankivell
Auto-Architect – The Autobiography of Gerald Palmer
Roads to Oblivion   Christopher Balfour

ACKNOWLEDGEMENTS

I would like to thank contributors to the Jowett Car Club library – club members, former employees and enthusiasts – whose photos, together with my own personal collection, have made this book possible. A special mention goes out to my son Ben, whose computer skills have helped me immensely. I would also like to thank my family; my wife Jane, and children Jonathan and Jessica for their invaluable support during the compilation of this book, and throughout my involvement with Jowetts since 1984. I also thank the many friends I have made within the Jowett Car Club and affiliated clubs both at home and overseas, who make Jowetteering such a pleasure. C.H. Wood Ltd of 500 Leeds Road, Bradford have kindly given permission to use photographs from their collection of historic Jowett images. Copies of these are available from the address above.

CONTENTS

William Jowett

Benjamin Jowett

William and Benjamin Jowett, the Bradford brothers who set up the Jowett Motor Manufacturing Company in 1901.

A nice example of a Jowett Bradford demonstrator, owned by the Sheffield agents, Hallamshire Tyre & Motor Co. Ltd. The Bradford was Jowett's most profitable model after the Second World War, popular both as a private and as a light goods vehicle. Note the selection of different petrols for sale.

One

THE BEGINNING
OF A DREAM

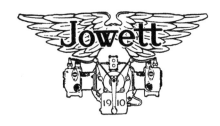

Wilfred Jowett was a blacksmith, living and working at 87 Kensington Street, Girlington, Bradford. He had five children, Ruth Elizabeth (1875), Benjamin (1877), Joseph (1878), William (1880), and Beatrice Anne (1882). Wilfred's wife, Sarah, died in 1888, and Joseph died in infancy prior to her. Benjamin and William found themselves helping their father in the blacksmith's business, and were involved in repairs of anything mechanical. By the late 1890s they were working for him full time.

Benjamin and William had a dream of building their own car and became more and more involved in engine repairs. In 1901, Benjamin, William and Ruth left the family home to set up their own business – The Jowett Motor Manufacturing Company – based in Church Street, Bradford, with each investing £30. For some time the brothers had wanted to produce a better engine with less noise and vibration to be used in existing cars, in particular the De Dion and Aster. The engine they designed was a 55° V-twin water-cooled arrangement, which was remarkably smooth running and reliable. This was used in several cars, and was also sold as a stationary engine.

During this period, Arthur Lamb joined the business by buying Ruth's third share at £60, as the concern was valued at £180. He was a well-known cycle engineer, but was to undertake clerical duties, which allowed the brothers to concentrate on engine design. The business moved in 1904 to larger premises nearby at Back Burlington Street, off Manningham Lane, Bradford, as access was difficult and space was at a premium in the old premises. The new property was ideal for car repairs, etc., as it was a single storey property with an area of over 400 square feet.

Experiments continued with engine design and, in 1904, a three-cylinder, in-line, air-cooled engine with many advanced features, including overhead valves, was produced. This did not reach the production stage, due to problems with balancing the crankshaft. It was also estimated that it would be too expensive to manufacture.

The next engine to appear, in 1905, is the one known and loved by Jowett enthusiasts the world over. The new twin-cylinder, horizontally-opposed unit was rated at 6.4 horsepower. This engine was used in the first prototype car built by the brothers and registered AK 494 on 14 February 1906. The car was used and tested by the brothers extensively over the next four years. Other engines were also tested but the engine in the prototype proved to be the best. A local coachbuilder called Ryder built the body for the car, which was described by the Jowett brothers as the world's first purpose-built 'light car', a justifiable claim, as it weighed just 6cwt. The general engineering side of the business continued to grow rapidly and, by 1907, the Back Burlington

Street premises had been outgrown. Fortunately, the brothers were offered the premises at Grosvenor Road, Bradford, giving them the much-needed space to expand their business.

The brothers continued with general engineering work while testing the car. In 1908 they were approached by Alfred Scott to produce six motorcycles to his design. This contract caused many problems, as Scott was constantly changing the specification. Despite this, they completed the project in January 1909. The motorcycle had a two-stroke, air-cooled engine of 333cc, capable of 43mph and 85mpg. All six were built at the Grosvenor Road works. However, the contract did not go according to plan, as the brothers had expected to then produce these motorcycles, whereas Scott had set up his own company to build them in Saltaire, near Bradford. He also left the brothers to dispose of the six machines! This proved to be an expensive exercise for the brothers, but it did give them valuable experience in solving mechanical problems, and the ways of business.

By early 1910, the brothers felt the car had been tested enough, and that it should be put into production. The first batch of twelve cars was built between 1910 and 1911 and still retained the tiller steering. The design now had a small, flat boot, unlike the prototype whose body ended at the rear of the seat. At the time the first car was produced, an Englishman living in Cape Town, South Africa, advertised that he would buy a car capable of climbing Table-Top Mountain in one go, as all others failed to do so due to boiling over or not being up to it! He challenged any motor manufacturer to supply a car on a sale or return basis. Needless to say the brothers were up to the challenge and shipped the car over. The car was never returned, so it must have been worthy of the task! The brothers soon realised that with the low RAC rating of 6.4hp, it was causing public resistance, as people at the time felt a car was inferior if it was too small. In true Yorkshire fashion the brothers altered the advertising to read a rating of 8hp, even though no alteration was made to the car at all. This did the trick as sales then took off and never looked back!

Between 1912 and 1916 a further thirty-six cars were built, with many modifications and improvements being built-in, the most noticeable of which was the introduction of a steering wheel in 1914 in place of the tiller steering. The Jowett was probably the last production car to switch from tiller to steering wheel – us Yorkshire Folk don't like change!

By the end of 1916 all car production ceased, and the factory was used during the remainder of the war for producing armaments and manufacturing machinery. Small components were also produced for the Rolls-Royce 'Eagle' aero engine as well as brake shoes for Crossley and Albion Motors. This was a profitable time for the brothers, but was also a great help to the war effort. After the war, Jowett had little work, but were not allowed to lay men off due to government policy. This created a loss for the business in 1919 but, owing to the hard work of Arthur Lamb, it received a refund of 'Excess Profits Duty.' This was on the proviso that the money was used to upgrade the Grosvenor Road property, or to build a new factory.

Benjamin wanted to form a new company and to build a new purpose-built factory for the reinstatement of car production. A worked-out stone quarry was found on the northern outskirts of Bradford at an area known as Five Lane Ends, Idle. It was looked at and purchased by the brothers. This site, at the time, was in open country, but was on the tram link into Bradford. They were able to buy the site for £100 due to its perceived limited potential. The brothers then sold tipping rights to Bradford Council, in effect paying nothing for the site at all!

Things moved quickly and the new factory was built during 1919 and measured 100ft by 150ft. The Back Burlington Street property was sold and a new company, 'Jowett Cars Ltd', was registered on 30 June 1919. Light engineering work was still carried out at the Grosvenor Road works by the Jowett Motor Manufacturing Company Ltd, and this would become the main sales outlet for Jowett Cars Ltd. At the end of 1919, lathes, milling machines and other equipment were shipped from the Grosvenor Road works to the new premises. Also, many items of Government surplus machinery were purchased including brazing hearths, blacksmiths forge, and a case-hardening plant. Everything was ready for car production by early 1920, and Jowett Cars Ltd was born!

The experimental three cylinder engine of 1904. It was air-cooled, and had a cubic capacity of 950cc. It was designed to run in either direction, so could be fitted to a car with no gearbox. There were problems in balancing the crankshaft, and it was expensive to produce, so did not go into production. One example was used in the Jowett works as an air compressor for thirty-five years!

The little engine with the big pull! This basic twin cylinder, horizontal-opposed engine was used from the first production car, right up to the closure of Jowett cars in 1953, when it was still being used in the Bradford van. This gave it a production life of forty-seven years, a feat which was noted in the Guinness Book of Records!

The Jowett brothers had tested the prototype car for four years and decided to start car production in 1910. This is one of the first batch of twelve cars of c.1912, with Ryder bodywork.

This is a 1913 example and was the twelfth car to be made by Jowetts. It has been owned for many years by the Peter Black Car Museum in Keighley, but has recently been sold into private hands, and is being fully restored again.

This is a 1916 example and was the forty-third car built. This picture was taken in the 1960s. The car is now owned by Mike Koch-Osborne, the grandson of William Jowett.

The two-seater of 1921, now with steering wheel, and Humbouldt-type coachwork. A questionnaire was sent to all customers in late 1913 to see if the tiller should be altered to a steering wheel, even though most other car producers had switched a decade prior to this! The Jowett brothers introduced the wheel in 1914, even though the questionnaire result was in favour of the tiller – Yorkshire-folk don't like change!

An informal gathering of Jowetts c.1922. Sadly the location and details are not known.

An early postcard of Kirkby Lonsdale, with a c.1916 Jowett parked in front of the Waverly Temperance Hall. It is thought the car belonged to the photographer. It looks like there is another Jowett behind the motor cycle and sidecar down the street.

Two

THE ROARING
TWENTIES

By April of 1920 car production had started again, the cars were exactly the same as the 1916 models, with bodies by Humbouldt of Bradford, which were of a very high standard. In late 1920 Humbouldt ceased trading. The brothers were quick off the mark and were able to purchase all the relevant tools, jigs, and timber stocks to enable the bodies to be built in-house. Also in late 1920 the new Road Traffic Act was introduced, which taxed cars at £1 per year per horsepower. The old advertising of the 8hp car was altered to 7hp, so anybody reading the adverts would think the engine size had been reduced by 1hp when in fact it had been increased by over 0.6hp! This allowed the Jowett to be taxed at £7 per annum, and was another excellent example of Yorkshire logic to solve a problem! 1920 proved to be a good year for Jowett, with them producing 100 cars, more than twice the pre-war production total.

The demand for the cars continued to grow and 1921 proved to be a busy year. More staff were taken on, and the workforce grew to 150. It was still very much a family concern, as so many of the new employees were friends or relations of existing staff, but all were committed to the job they were doing. During 1921 it was felt that the old Humbouldt body, used from the start of production was becoming rather out-dated. A new body was therefore designed with a higher radiator grille and bonnet, with a higher tail, as well as a dickey seat for two installed. This design was ready for, and displayed at, the White City Motor Show in November 1921. This was to be the 1922 model, and a commercial was also introduced at the same time. This was another very good year for Jowett with 249 cars being produced, at the rate of ten a week by the year-end. The new model was well liked and proved to be a great success, with growing sales of commercials also. The grand total of chassis production for 1922 rose again dramatically to 492.

There was a real camaraderie between Jowett owners, and even the owner's handbook suggested that it was expected to wave to an oncoming Jowett owner! During 1922 and 1923 various Jowett clubs were set up around the country, for social and motoring events. Car sales in 1923 continued to grow rapidly, the long-four (four-seater) model, on a longer chassis was introduced, and the short-two and commercial continued to be built. Total sales in 1923 were 1,047, and up to 1,853 in 1924 increasing to 2,223 in 1925.

Jowett advertising was always interesting and different to other car manufacturers, with slogans building up through the 1920s. As early as 1912 Jowett were advertising how cheap it was to run and maintain a Jowett and how reliable they were: 'one penny a mile'. During the 1920s it was in a class of its own with other gems such as: 'The little engine with the big pull'; 'Any road any load'; 'If you want to go where Jowetts won't – you'll need a crane'; 'Jowetts never wear-out, they are left to your next of kin'; 'The pull of an elephant, the appetite of a canary, the docility of a lamb'; or 'The Seven that passeth the Seventeen like a Seventy!' These slogans and adverts were produced by Harry Mitchell and Gladney Haigh. This type of advertising was to continue right through the 1920s and 1930s. Adverts in *The Motor*, *Autocar* and *The Light Car And Cyclecar* took the form of a full page of text in a floral border, normally with very little technical details of the car, and in most cases, no picture of one! Instead they would extol the virtues of Yorkshire, and motoring in general. This policy did work, as sales continued to grow. In fact it is said that many people bought the motoring publications just to read the Jowett adverts!

Owners were also encouraged to write in to Jowett to say what they had done with their car, and how economical and reliable it was. These testimonials were regularly published in little booklets and sent with other sales literature to prospective buyers. A typical example regarding economy published in 1924 read: 'There's a Jowett owner in the south of England who regularly obtains 60mpg of petrol. He vows that with a good gramophone horn fixed to his Carburetter intake, he could run on the smell of a London omnibus. So there is truth in the "Canary Appetite". But don't put seeds in the tank!' A typical Jowett advert from 1924 read: 'We hate to think a time may come when Jowett Cars will be bought, used, and sold like any other. Up to the present, and for nearly 20 years, they have been bought discerningly, but with ever-increasing amazement at their capacity for unremitting labour, and when sold, parted with as a trusted friend, with very sincere regret. When you buy one, it will endear itself to you, more and more, every mile you run it.'

A good example of clever Jowett publicity came about in 1924, when a new sewerage system was to be opened between Bradford and Esholt. The mayoral party wanted to lay the last brick, which required a journey of three miles down a narrow sewerage tunnel in cars. Needless to say the Jowett brothers jumped at the challenge and laid on four long-fours for the ceremony, which made the trip without incident. To boost overseas sales, two booklets were produced in 1925 giving details of expeditions carried out by Jowetts. The first was called *Where there's a way the Jowett will go* and described the 840 mile trip from Alexandria across the Libyan desert to Siwa, and back via Gara. The party included one 1924 Jowett short-two and three American cars. The little Jowett proved to be far superior to the other cars as it never became bogged down, and never needed to be pushed or towed. It also averaged 47mpg, three times better than the American cars. The second booklet gave details of a similar trip, with similar results, from Cairo to Siwa and back on a previously unused route covering 930 miles. Jowetts made quite a name for themselves in Egypt, and there was an active Jowett club based in Cairo!

In 1926 Frank Gray, the former MP for Oxford, must have been a godsend to Jowett Cars, as he had been complaining that British cars were not suitable for colonial use. He challenged the British Motor Industry to provide him with cars which would be capable of crossing Africa, carrying all their own petrol, water and provisions, as most of the journey would have to be covered without any outside help. This was due to the fact that there were virtually no roads or services in Africa at that time. Needless to say there were no offers forthcoming from other car manufacturers, as such a trip had never been attempted before. The Jowett brothers, however, realised the free advertising potential such a trip would create and agreed to sell Gray two cars. Being two shrewd Yorkshiremen, they said they would repay Gray the cost of the two cars if the trip was successful. This way Gray would also have a financial interest in completing the challenge!

Gray would drive one car and Jack Sawyer, a wealthy neighbour of Gray, would drive the other. It should be noted that Gray could not drive but would learn on the trip! Sawyer travelled to the Jowett factory and chose two long chassis two-seaters, and asked for a lorry flat-back to be fitted in place of the dickey-seat. He also required tow bars fitting to the cars, as they would both have to tow trailers loaded with provisions, etc. At a press conference prior to the departure of the cars for Africa, a reporter asked Ben Jowett if the cars had any real chance of completing such an arduous journey, he replied 'Wait and see'. Gladney Haigh heard this and ordered the two cars to be sign-written on each side with one called 'Wait' and the other 'See'!

The crossing was to be from Lagos in the west to Massawa in the east, and most of the route would be across open country and desert as there were few roads of any kind. While in Lagos,

14

Gray and Sawyer enlisted the help of two locals, one as a cook and one called Bismark as the mechanic. Bismark agreed to make the crossing, provided that they would take him to England for an extended holiday afterwards. Gray and Sawyer agreed to this request. Prior to the start of journey the local dignitaries invited them to dinner, none of whom gave them a chance of success. The expedition set off from Lagos on 16 March. The going was good and they covered almost 300 miles in the first two days. They reached Kano to meet up with the trailers that had been sent ahead of them. These were loaded with petrol, water, tyres and provisions. Gray and Sawyer were now on their own for the next 1,600 miles of mainly desert until they reached El Obeid. During this part of the journey the terrain was very difficult and they were lucky to cover 100 miles per fourteen-hour day. The heat was so great that the bodywork of the cars could not be touched during the day without causing burns. There were various problems *en route* but neither car had any major mechanical problems, and the expedition reached Massawa sixty days later on 14 May. This represented only forty-nine driving days, as eleven days were rest days. In all, a total of 3,800 miles were covered and they even had time to rescue a slave girl and transport her 120 miles with them to a district judge and safety!

Needless to say, the press loved this story of British endeavour and Jowett Cars received vast amounts of publicity. Gray and Sawyer were true to their word and brought Bismark back to England with them. Bismark and the two cars went on tour, calling at Jowett agents to allow the cars to be inspected. This also generated a great deal of publicity in local papers. The cars were later used as factory transport at Jowetts for several years delivering spare parts to agents. The fate of these two historic vehicles is not known.

1926 also saw the introduction of the first Jowett saloon. It was known as the 'Greenhouse Saloon' as it had very large windows! Sales continued to grow with a peak of 3,474 reached in 1927. Sales were so good that a night shift was now in operation. Another advertising opportunity for Jowett occurred with the sale of twenty-seven long-fours to the Metropolitan Police in 1927. These cars were driven down to London in convoy calling in at Jowett agents on the way down. They were a success, as this advert from 3 August 1928 confirms: 'They've made good because they're made good – Scotland Yard after 18 months' experience with their fleet of Jowetts, have just honoured us with another large order. Jowett cars have the best put into them, and Jowett owners get the best out of them. They were never cheaper, never more reliable, and Jowett owner's expenses are comparatively negligible. Follow the lead of the experts and buy one today!'

Jowetts were also tested by the War Department in 1927, but no orders were placed at that time. A sports model and a fabric saloon followed in 1928, and the 'Black Prince' saloon in 1929. This was a milestone model according to Jowett advertising: 'Last year the motoring cognoscenti declared we had reached finality in design. We were rather of the same opinion ourselves, until the introduction of the Black Prince fabric saloon in the early part of the season proved that the pinnacle had not hitherto been attained!'

This had been a golden decade for the Jowett brothers. Their product had been well received, and had built-up a strong loyal following with car owners.

The following shots are of the 1922 short-two, with bodywork by Jowett. The model would prove to be a real success in the 1920s.

The tail of the car would open to reveal a dickey-seat, which would seat two people, ideally children! To get in, the first step was the running board, then onto the step on the rear mudguard.

The side view of the 1922 short-two.

Three-quarter rear view with hood erected.

The 1922 Jowett was displayed at the October 1921 Motor Show. It was decided to demonstrate the new car to the press. William Jowett made many assents of Spion Kop, which was a mound or slagheap about 100ft high on the Jowett factory site. The first part of the climb was 1 in $2\frac{1}{2}$, and the second part was 1 in $3\frac{1}{2}$. This shot shows William on his own, but on some ascents there were four people on board!

'This Freedom!' was an early Jowett slogan, as it brought car ownership to many working people for the first time. This lovely shot of a 1923 short-two, registered KU 2736, was taken in the Dales. Note the two youngsters in the dickey-seat.

A nice view of a 1925 short-two, registered KU 3679 in Matlock Bath. KU was a Bradford registration.

Another photograph of KU 3679, this time on a family trip to Burton-in-Lonsdale.

Frank Wood worked for the publicity department, and would arrange to take new cars out on photographic sessions. The girls featured were models he hired to add some glamour to the pictures. Frank would normally get into the pictures also! The following four photos were taken in 1926. Here he is in a short-two.

Frank waves from a long-four.

Frank at the wheel of a short-four chummy.

Frank assists a young lady into his long-four!

Light commercial vehicles were of great importance to Jowett, and were built throughout Jowett's history. This is a 1925 example, built on the short-two chassis and using the same mechanicals.

Jowetts being put through their paces on the Jowett test course, c.1927. Not a route for the faint hearted!

The 'wait and see' cars of 1926, which crossed Africa in forty-nine running days. The first picture shows Frank Gray and Jack Sawyer en route.

The two cars arrive home to Bradford in triumph, note 'Bismark' on the back of 'wait.' He was the native mechanic employed by Gray and Sawyer. He agreed to go provided he was given an extended holiday in Britain after the crossing! As can be seen, they were men of their word as Bismark accompanied the cars on a publicity trip round Jowett agents.

Other examples of Jowett's ability to obtain good free publicity are shown in these two photos. The first shows Alderman R. Johnson (chairman of the sewerage committee) and members of the Bradford Corporation, who motored three miles underground in 1924 to perform the opening ceremony for the extension of the sewage system from Esholt to Frizinghall, Bradford.

In 1926 the Metropolitan Police placed an order for twenty-seven long-fours. They were all fitted with Humfrey-Sandberg free-wheel mechanisms, allowing the cars to free-wheel automatically on the level and downhill. This novel mechanism was said to improve petrol consumption by 15-20%. All the cars had boards attached to them saying they were for the Metropolitan Police. They were driven down in convoy, stopping at Jowett agents *en route*.

The 1926 Greenhouse saloon – thus called due to the area of glass – also had an opening windscreen.

This charming photo shows M. Winterbottom in his father's long-four. The passenger front tyre seems to have had better days, but there were no worries about the MOT in those happy days!

Above and below: J.J. Hall ran a Jowett sports model, which he drove happily in excess of 60mph. He contacted the factory, as he noticed that an Amilcar held the international Class G twelve-hour speed record at 54.24mph. He was sure a Jowett could beat this, so Horace Grimley of the experimental department was told to prepare a car for the event. The car was similar to the sports model but had been lightened, had no mudguards, and was built on a narrower frame. During the attempt in August 1928, they had three blown gaskets but managed to scrape in at an average of 54.86mph! Horace Grimley is seated in the car, with J.J. Hall standing.

Caravanning 1920s style. It would appear to be a rather heavy load for this *c.*1926 short two, but Jowetts had no problems pulling such loads.

A proud father out in his *c.*1927 Long-Four Chummy. It would be a full load with the four children once his wife got back into the car after taking the photo!

The Ministry of Defence tested Jowetts in 1927. This one is being put to the test on very rough terrain.

Gladney Haigh shows Lord Harewood the military Jowett. Frank Gray is also looking on (second from left), but this is not a 'wait and see' car.

'Yes, you must go to the ball Cinders', and what better way to go than in a Jowett! Note the rather nice lady radiator mascot.

The long-four with hood up and side screens in place.

Jowetts were very popular in reliability runs, a successful driver being W.S. Canney, who entered the Yorkshire Evening News £200 trails in 1927, 1928, 1929 and 1930. This photo was near the start of the 1930 event. The co-driver was Teddy Gascoyne, manager of the Jowett service department.

Near the end of the 1930 Evening News trail.

In 1928 Jowett produced fourteen sports models. The record-breaking car of J.J. Hall was based on this design. This is a publicity shot of the car.

John Frost raced this example in the 1930s. Registered RK 8538, it was painted yellow, with red mudguards and wheels.

Miss Victoria Worsley drove her sports model, registered KW 3400, in over thirty events during 1928 and 1929, with much success. This event was the JCC High Speed Trial of 6 July 1929, where she won a silver medal.

Victoria took part in hill-climbs, reliability runs, speed trials and dirt races, as well as sand racing. This event is possibly the Skegness Speed Trials of 5 July 1928.

Three

THE 1930S – DEPRESSION YEARS

The 1930s would prove to be a much more difficult time for the Jowett brothers. Early in 1930 there was a dramatic fall in the sales of cars as a whole due to the growing economic depression which was sweeping across the country and, indeed, the world. There had been lengthy waiting lists for most models in previous years, but by the summer of 1930 cars were being stockpiled. On 3 September there was a huge fire which swept through the factory and gutted most of the complex. This destroyed cars, equipment, stock, raw materials and production lines. The heat was so great that the stock of aluminium sheet melted into a solid lump. Over the next few days anything which was salvageable was removed and stored.

Much of the factory equipment was old, so when the insurance claim was settled it was for much less than the brothers had hoped for. They did wonder whether to continue, or not, but then decided to rebuild as soon as possible. This was still very much a family concern, and this trauma brought the workforce together even more than before. Incredibly, by mid November the new buildings had been rebuilt and one or two new commercials had been built. A few new 1931 models had been built prior to the fire and had been stored in the finished car bay, and only suffered smoke damage. The best six of these cars were refurbished while the rebuilding work continued. These cars were then placed on the show stand at Earls Court to let people know Jowett were still in business!

It would be a full year before full production would be achieved again. During this period the proportion of commercial vehicles increased to almost 50% due to two main reasons – firstly they were easier to make when the factory was still being refurbished, as they were more basic in design and not finished off to the same standard as a car and, secondly, the commercial vehicle market was not as depressed as the private car market so it made good economic sense to produce more. In view of the traumatic year the brothers had, it is amazing that as many as 2,603 vehicles were produced in 1930 and 1,706 in 1931.

Another source of income for Jowett during these difficult times was the sale of engines to other manufacturers. Karrier Motors of Huddersfield who, in conjunction with the LMS Railway, were trying to produce a mechanical replacement for short deliveries, which were at that time undertaken by horse and cart. They announced the Karrier Cob tractor unit in late 1930. This was a three-wheeled tractor unit with a simple coupling device to use with existing trailers. The Cob was powered by the Jowett engine and had a three speed gearbox. It was capable of carrying a three ton load at 18mph, and could start when fully laiden on a 1 in 8 gradient. By the end of 1931 all of the big four railway companies – LMS, GWR, LNER and SR –

were using them. In total over forty were in service. This was a short-term venture, as Karrier switched to a four-cylinder Coventry Climax engine in 1932.

A similar opportunity presented itself when Jowett were contacted in 1934 by Bristol Tractors Ltd of London. They had designed a small crawler tractor for use in market gardens and orchards. It was powered by an two-cylinder, air-cooled Anzani engine, which had proved not to be strong enough, and Bristol contacted Jowett for help. A number of engines were provided, but Bristol was not able to pay as it was in financial difficulties. William realised there were few competitors so invested in the company to stop them from going into liquidation, and became its managing director. The whole operation was then transferred to a disused property near the Jowett works in Bradford. Jowett engines were fitted to the tractors brought up from London. Engines made by Jowett continued to be used in these tractors right through the 1930s.

Production at the Jowett factory continued at a similar rate during 1932, then in 1933 the Kestrel was introduced. It was a four-light saloon with a new body styling, but with identical running gear to the previous models. The Focus (long-two with dickey) was renamed the Flying Fox. Other models, such as the Weasel and Curlew, together with commercials, continued to be produced during the 1930s and all were powered by the same twin cylinder engine. Then in 1935 there was something of a departure for Jowett with the introduction of a flat-four engine fitted with twin Zenith carburettors and with a 10hp rating. It was installed in the new Jason and Jupiter saloons. Both cars were identical in appearance but the Jason was the de-luxe version and had very plush trim while the Jupiter was the standard model. At this time there was a vogue for streamlined cars and these two models were an attempt to catch the mood of the day. The cars were displayed at Earls Court in October 1935. They had a very steep raked radiator and steep sloped rear. Sadly this model appeared too futuristic for most Jowett buyers, and sales were very poor. The brothers realised they had made a mistake with the styling, and within four months replacement models were introduced. The Plover replaced the Jupiter and had a new radiator grille and bonnet but with the rear of a Kestrel. The Peregrine replaced the Jason. This had the same front end as the Plover, but retained the Jason rear. In all only 105 Jasons and thirty Jupiters were produced and the two replacement models were also short-lived, only being produced in 1936 with a total production figure for the two of 166.

This was an expensive exercise for Jowett but the new four-cylinder engine had been well received. Reg. Korner designed a new body style for the 1937 saloons – the 8hp model using the two cylinder engine and the 10hp model using the four cylinder engine. These were identical in appearance and proved much more acceptable to Jowett owners. Both these models were very popular and sold in large numbers up to the outbreak of the Second World War. The commercial range also used both engine sizes, but the 10hp was used only in relatively small numbers. A revised design had been shown at the 1939 Motor Show, with an extended boot, which was to be the new model for 1940, but only 113 of these had been produced prior to hostilities stopping car production.

With the problems faced in the mid-1930s the brothers had differing views on how the business should go forward. There had been mounting pressure to float the business to raise capital for future development programmes. William was in favour of the flotation but Benjamin certainly was not. The flotation did however go ahead on 6 March 1935, much to Benjamin's annoyance, and Jowett became a public company. The two brothers were the majority shareholders. Benjamin would leave the company the same year. Various other problems and a decline in profits followed as basically the models were becoming outdated. They were no longer cheap since the plant and working practices were also in need of updating.

It was in this climate that in 1939 William also retired from the company. Both brothers retained their shares, which were held in trust, allowing them to retain financial control of the company. Charles Callcot-Reilly succeeded William as managing director. With the outbreak of the war the brothers sanctioned war production, and the cessation of car manufacture.

On 3 September 1930 there was a disastrous fire at the factory, which left most of the buildings, equipment, stock and cars in ruins. This picture was taken on 6 September.

Salvage work and reconstruction on 12 October 1930.

Reconstruction continues – 25 October 1930.

30 November 1930. Incredibly, the new buildings were up, with roofs on in just two months after the fire started. There was even time to renovate some damaged cars for the October Motor Show.

A 1930 van on test by Jowett using trade plates. The spare wheel is located on the roof, allowing more space in the van. Not much fun for the driver with a puncture on a wet day!

Another view of the 1930 van on test.

The 1931 range of commercials were advertised under the slogan 'Transport Economy. The Little Engine With The Big Pull.' It was still possible to advertise the total running costs at a penny a mile.

The JOWETT STANDARD VAN

The standard 1931 van, which was bought by fleet buyers such as Bassetts Liquorice Allsorts, Players and Andrews Liver Salts.

The JOWETT LORRY

The 1931 Lorry.

The 1931 Covered Lorry, which was very popular with milkmen.

The 1931 Dual-Purpose Saloon. Was this the world's first hatchback?

In 1931 Jowett produced a fleet of thirty Reconnaissance Vehicles for the Artillery Transport Company of York, for use by the Territorial Army. They were used for cross-country telephone cable-laying. This publicity shot shows eleven of the fleet.

A close-up publicity shot of one of the Reconnaissance Vehicles. The official caption reads: 'Complete with telephone cable gear etc. This car has been specially produced for the carrying of battery staffs of mechanised units and can transverse any class of country bog and ditch, with seating capacity of four men and equipment.'

A Kestrel receiving a final polish in the finished car shop in 1933.

A selection of Long Saloons being checked and prepared in the finished car shop in 1933. Note how smart the workers are in view of the work they are doing.

A well known publicity shot of a 1931 covered lorry, being used as a milk delivery van by G.H.I.S. Ltd of Bradford

Eric Lang, a gentleman's outfitter, stands proudly with his 1933 van, bought from the Jowett agents in Kendal. It was used extensively by him on his large sales area covering much of the Lake District.

Camping 1930s style. This 1933 Kestrel is heading off into the countryside towing a Rice folding caravan.

A petrol station 1930s style. A *c.*1930 Fabric Short Saloon filling up with Double Shell at 1s 8d a gallon.

The Jowett twin-cylinder engine was used by Karrier Motors from late 1930 to early 1932 in their Mechanical Horse. This proved very popular with railway companies, for whom it was designed.

The Jowett engine was also used in the Bristol Crawler from 1934. This one used the twin-cylinder engine, but later four-cylinder engines were also used.

An unlikely combination! The Honourable Mrs Bruce with her Fairey Fox, probably the highest-powered privately-owned aircraft in the country in 1933, together with her Jowett Kestrel. She completed the 1933 RAC 1,000 mile trial. Later in the year she and her husband drove the car continuously for seventy-two hours on the Montlhery Track near Paris, covering 2,775 miles at an average speed of 38.54mph.

A 1935 commercial used by the Jowett agents, Buckrose Motors Ltd of Bridlington, advertising Exide batteries.

Above and below: What appear to be two period publicity shots using a 1916 Jowett. The first is with a 1931 saloon. The second with a 1933 Kestrel.

The first Jowett to depart from the flat-twin engine was the 10hp flat-four Jason saloon, which was the deluxe version, and the Jupiter saloon, the standard model. Sadly these were not a success, as the public did not like the steeply raked radiator grill. In all only 105 Jasons, and thirty Jupiters were built.

In 1933 and 1934 Jowett built and tested two experimental cars called 'La Roche' – this was an anagram of chorale, a slow and dignified hymn. It had a conventional in-line four cylinder engine. The bonnet badge only showed 'La Roche', with no reference to Jowett. Testing was mainly done at night, as Jowetts did not want the public to think they were planning to discontinue the 'little engine with the big pull'. The project was abandoned in 1934. This is the second car, which was sold to Rupert Lindly, the company auditor.

The 1932 Kingfisher long wheelbase saloon, one of 1,067 built that year, photographed at a club rally in 1981.

The 1934 Kestrel, pictured at a club rally at Malvern in 1979.

Large numbers of Jowett chassis were exported to Australia in the 1930s, and were bodied locally (Bradfords were later sent too). These four photos were taken in Melbourne in the late 1930s and show the different available models starting with the 8hp Tourer.

The 8hp Utility.

The 10hp Saloon.

The 10hp Tourer.

A Jowett 8 saloon. The sales booklets stated: 'It is doubtful whether any other manufacturer of a light car has so many instances recorded of cars that have gone on, year in and year out, giving as good service after ten years, as when they were first bought. What is the reason? It isn't just that they are made so thundering well by Yorkshiremen up in Yorkshire. The reason is to be found in the unique horizontally opposed Jowett design.'

The 1938 Jowett 10hp saloon. The 1938 sales booklet says, 'to the motorist who wants performance, but who studies economy, we say "balances power" makes the 10hp opposed four, just the car for you.' Petrol consumption was said to be thirty-two miles per gallon.

From 1938 the 10hp engine was also available in commercial vehicles, together with the 8hp models. They were made in much smaller numbers. In 1938, sixty-nine vans and nineteen lorries were made. In 1939, ninety-six vans and forty-one lorries while in 1940, thirty-six vans and four lorries left the factory. Sadly no examples are known to survive. This is the 1938 10hp van.

The 1938 10hp lorry.

At the 1939 motor show, the 1940 saloon was displayed. It had an extended boot and only 113 of these cars were built in early 1940, before war production started. This example was photographed in the 1960s. It was thought that no examples of this model had survived, but recently one was found in a derelict state and is now being restored.

The same car from the rear.

Four

THE WAR YEARS

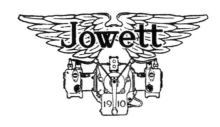

As in the First World War, this period proved to be a time of great expansion for Jowett and many other manufacturing companies. The factory was completely redesigned and expanded, the workforce also grew four-fold to 2,000, and most of the new workers were women. New workshops were opened to produce field guns. Other items produced included capstan lathes, stationary engines, ammunition from bullets to 25-pounder shells and airplane parts. Reg Korner of the design team also designed a mobile mortar launcher, which proved to be a great success, with over 600 being produced. The workforce were also involved in Home Guard duties and the factory also had its own fire service, with these duties being arranged by Horace Grimley.

Charles Calcott-Reilly was very much aware of the fact that when peace was announced, Jowett would need new models that could be exported all over the world. He knew that exporters would have priority for raw materials, and that funding would be needed to switch back from war work to car production. How this was achieved is described in later chapters.

Sadly, William and Benjamin were by now barely speaking to each other and their relationship was strained to say the least They did agree, however, that they did not want to fund the operation to restart car production at the end of the hostilities, and they therefore sold their shares in the company. The postwar Company would therefore have no direction from the Jowett brothers. Was this an omen for the future?

On a lighter note, Jowett produced a booklet in 1946 called *War Production Record* which detailed every item produced for the war effort and waxed-lyrical on how this was done:

> 'Then came the dark days of Munich and the start of the present war, when that fateful September of 1939 broke upon Europe, Jowett were peacefully turning out cars. We switched almost overnight to capstan lathes and aircraft components. When a man in the street saw the bombers roaring overhead and the guns go rumbling through the streets on their way to the ships, he could feel a certain glow of satisfaction and comfort in the knowledge that they kept him safe at home.
>
> When Jowett workers looked up and saw the same aircraft, they could feel proud that they had helped put them in the sky. With their own hands they helped to make hundreds of guns and millions of ammunition for them. Day in, day out, nights as well, the Jowett workers stood at their benches and made the countless thousands of parts for planes, ground equipment, guns, ammunition, tanks, motor engines and machine tools.

But now we see the results in Europe, the results of all this labour and the work of the men who used the weapons made by Jowetts. During those six long years, did Jowett workers ever wonder how much the factory was turning out, and what the value of the output amounted to? Did they ever think that if those fittings they stood and made had been faulty, their own sons and husbands and sweethearts might so easily have been killed? Perhaps they did, for the standard of their work never slipped.'

A 4.2in mobile mortar designed by Jowett cars on a gun carriage. This was lighter than the previous model approved by the War Office, and was also able to be produced at half the cost. The new strengthened baseplate was also a Jowett design, and they supplied 90% of all the army's requirements. They also produced hundreds of Sherman Flail mechanisms to fit onto tanks to help in the clearance of mines. Jowett produced a booklet in 1946 entitled the *War Production Record*, which detailed every item produced for the war effort. This included aero engine components, air ministry ground equipment, ammunition, armoured stationary engines and generators, machine tools, and other miscellaneous items, to a grand total of £4,777,804.

Five

THE BRADFORD

From as early as 1942 Charles Calcott-Reilly was planning the production of the new post-war car models and he knew these would need to be all-new and suitable for the world market. The volume of production was also expected to rise dramatically, due to the war-period expansion. The new car model was to be the Javelin, which is covered in the next chapter, but Jowett needed something to sell while work on the Javelin took place.

At the end of the war in 1945 Britain had massive debts and the Government watchword was 'export, export, export!' All raw materials were in very short supply, and car manufacturers had to export to qualify for these materials. The Javelin was not expected to be available for at least two years. It was therefore agreed to re-vamp the pre-war 8hp commercial as it was simple to build and it was expected to sell well. A new radiator grille was designed and it was to be known as the Bradford, so buyers would know were it had been built. A decision had already been taken that bodies would no longer be built in-house and they would be supplied by Briggs of Dagenham. Briggs also had a plant at Doncaster, which would be ideal for Jowetts as it was only thirty miles away from the Jowett factory. A sales booklet on the Bradford was available by December 1945 and, by early 1946, production was underway. The first design was known as the 'CA', which was the basic pre-war design. It was originally only available in van form, but by June a lorry version was introduced.

Demand was high and waiting lists grew. By November 1947 over 5,000 had been produced. At this time the second-series Bradford or 'CB' was introduced. It was still the same basic design, with the 1005cc engine. The engine now had a down-draught carburettor fitted, and a belt-driven dynamo, which allowed a cooling fan to be fitted for the first time. This was an optional extra but proved popular in export vehicles going to hotter countries such as Australia, Brazil and Uruguay. This model continued in production into late 1949 when the third series Bradford or 'CC' was introduced. The engine was completely redesigned with the RAC rating of 19bhp increasing to 25bhp. The electrics were finally also altered from 6 to 12 volts. This was not before time as Jowett received many complaints from owners due to the inadequacy of the 6 volt system. The vans literally ran out of power on dark, wet nights when both headlights and wipers were in operation!

The 'CC' model was really seen as the last stop-gap model prior to the introduction of the completely new range of car and commercials which were to be known as the 'CD' range. These had been expected to be ready by the October 1951 Motor Show. This range never went into production, so the 'CC' range continued up to the closure of the factory in 1953. By this time 38,241 Bradfords had been built. The engine design was basically the same as the first one designed by the brothers in the 1906 prototype car, a production run of over forty-seven years.

The Bradford was basically a pre-war design which was meant as a short-term model prior to the launch of the Javelin. It was a reliable, uncomplicated, frugal workhorse in the true Jowett brothers' tradition. It was loved by shopkeepers, small business owners and the general public alike. Ironically, this was the only profitable model Jowett produced in the post-war period and the profits generated from this model were used in the development and launch of the more glamorous Javelin and Jupiter.

The Bradford was rushed into production in 1946, and was a re-vamped, pre-war 8hp commercial. This early publicity shot shows what was planned for the front grille, but this was altered prior to it going into production.

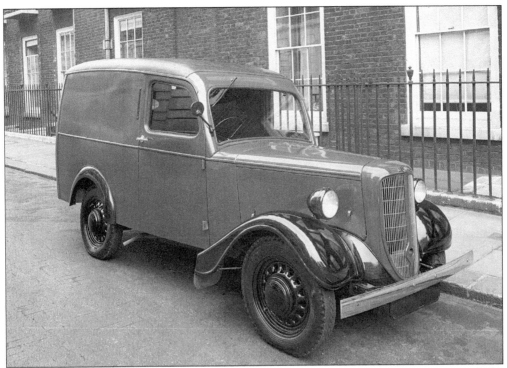

Publicity shot of a 'CC' Bradford of 1950. There was only one windscreen wiper and the passenger seat was also an extra!

Large numbers of rolling chassis were sold to enable the fitting of various body styles. This is a nice example of a 'Woodie' shooting-brake.

Hovis Bread also used Bradfords. This example is again in the style of a 'Woodie' shooting-brake.

Another interesting design, with raised roofline to allow better access for the rear passengers.

A Bradford at the end of the production line. Bradfords and Javelins were built side by side.

A finished Bradford leaving the factory on trade plates. Note the small oval rear windows.

The inside of the 'CKD' department at the Jowett factory, with a selection of utilities being attended to.

A well restored example of the Bradford lorry, taken at Beamish prior to the club's seventieth anniversary national rally, held there in May 1993.

A nice example of the Utility De Luxe with chrome headlamps and grille and a step by the drivers door.

The Bradford 'six light', i.e. three windows on each side, excluding door windows. This example is owned by the Bradford Industrial Museum, who have a good collection of Jowetts on display.

Several Bradfords were bodied as ice cream vans. This is one of only two known survivors.

This Bradford is owned by Wm Morrison supermarkets, which is fitting, as the former Jowett factory at Five Lane Ends, Idle, now has a supermarket owned by Morrison's on it. They bought and restored the van to acknowledge the historic nature of the site they now occupy. This van can be seen at classic car events most weekends during the season.

Six

THE JAVELIN

As mentioned in the last chapter, Charles Calcott-Reilly was planning for replacement models as early as 1942. He advertised for a new chief designer with a view to starting work on the new car straight way. One of the applicants was Gerald Palmer, a young engineering draughtsman who was working at MG at the time. It would appear that Gerald was not impressed with his trip up north to Bradford, and returned to MG. That could have been the end of the story, but Calcott-Reilly was most impressed with him! He made a special trip south to see Gerald as he was sure he was the man for the job. He offered Gerald a salary of £500 and a free hand to design the new car. This was an offer he could not refuse and he took the job!

Gerald was born in Rhodesia in 1911 and came to England, against his parents' advice, in 1927. He was apprenticed to Scammel lorries of Watford. In his spare time he worked on building his own sportscar which he called the Deroy. Gerald left Scammel in 1937 with a view of putting the car into production. Sadly the venture failed and only one car was produced, which Gerald kept for himself. It was not a complete disaster however, as the design work for the car impressed Cecil Kimber of MG. Gerald was placed in charge of the design team working on the YA saloon, and met Issigonis who was working on the car's suspension. It was when Gerald was working on this car that he was head-hunted by Jowett!

Gerald arrived in Bradford in September 1942 and was given his own office, which had a drawing board and very little else. His brief was simple – design a car that would appeal to both the home and overseas markets alike, with a cost ceiling of £500, the rest was down to him! He was soon at work experimenting on car and engine layouts, as he wanted to maximize the space inside the vehicle as much as possible. This found him looking at the idea of a flat-four engine placed well forward, above the front axle. His idea for the flat-four engine was put to the Jowett board, who loved the concept, as it was in true Jowett tradition! He was given the go-ahead to continue his experiments on the engine design and body styling. It is interesting to note that when Issigonis was working on the Morris Minor, he also preferred the idea of a flat-four engine, but was not given permission to proceed with it.

The engine design had a single, cast aluminium crankcase with the crankshaft running in two main bearings with a volume of 1184cc, but this was found to be too harsh, so other ideas were tested. The final design that was adopted for the car was, a vertically-split crankcase in die-cast aluminium held together by substantial tie bolts, using twin Zenith carburettors and having a volume of 1486cc. A four-speed gearbox with a column-mounted gearchange was used which allowed three abreast seating in the front, making the car a true six seater. The car also used front independent suspension (which Gerald first used in his

Deroy) but the rear had the more conventional beam axle and transverse torsion bars. The floor was flat, as the transmission ran below it, giving passengers maximum leg room.

The first prototype was completed in 1944 and was registered DKY 396 on 25 August. This car, and later two prototypes registered DKY 463 and EAK 771, were constantly tested through 1945 and into 1946. Regular meetings also took place during this period with Briggs Bodies, to work out the best production methods to make both Bradford and Javelin bodies. A large network of dealers and agents were being built up in readiness of the car being available. London showrooms were also acquired at 48 Albemarle Street. The first two prototypes were fitted with two piece windscreens, but EAK 771 was fitted with a single curved screen, which was developed by Triplex. This would be fitted into the production cars making the Javelin the first British car to be fitted with such glass.

EAK 771 was displayed to the public for the first time at the London Cavalcade of Motoring on 27 July 1946, and at a similar event in Edinburgh on 7 October. Needless to say the car was a sensation, and the motoring press could not wait to get their hands on one! Some reports did appear in the press to say the Javelin was coming, but the official launch date is regarded as May 1947, as this is when *Motor* and *Autocar*, etc., were able to test it. To say the Javelin was well received by the motoring press was an understatement – they loved it! Jowett's claim in its advertising, 'It's new right through', was true: it was the first all new car after the war, rather than just a revamped pre-war model like all it's competitors. It really was in a class of its own, with a top speed of 80mph with 30mpg, and acceleration to 60mph in twenty-three seconds. It was roomy with good handling, a real sports saloon. The sales slogan 'Take a good look when it passes you' was very apt, as very few other saloons could even approach this level of performance.

Production of the car was frustratingly slow due to the shortages of raw materials, and it was not until mid-1948 that cars were available to the public. Even then most went for export in order to qualify for more raw materials. By the end of 1948 only 1,558 Javelins had been built, but things improved in 1949 with 5,450 being constructed. During 1948 an interesting prototype was produced for Jowett by the Carlton Coachworks in London, being a drophead Javelin. This had the same front end and front bench seat, but with a long tapered tail and dickey seat, instead of rear seats. It was designed with export to hot overseas countries in mind, but removing the roof weakened the car, so it did not go into production.

Two well known Yorkshire rally drivers, Tommy Wise and Cuth Harrison, contacted Jowett to see if they would provide them with a Javelin to take part in the January 1949 Monte Carlo Rally. Jowett agreed to this request, as it was felt it would keep public interest in the car until it became more readily available. They made the stipulation, however, that the car's designer, Gerald Palmer, should also go on the car's international rally debut. Wise and Harrison agreed provided that Palmer did not drive; they felt he would be too gentle with his baby! The car had a trouble-free run and won the 1½ litre class, being placed fourteenth overall. There were also several other Javelins in the finishers.

In view of the Monte success Jowett decided to enter a Javelin in the Spa 24-hour race in Belgium in June 1949. The company felt that a good result here would enhance sales in Europe. Anthony Hume of Motor and Tom Wisdom of the Daily Herald would drive the car this time. The car was driven to the event, along with another car carrying spares, and it was then prepared and raced for twenty-four hours. It was entered in the 2 litre class, which it won with ease; in fact it was lapping faster than the 4 litre touring cars. It is interesting to note that there was a 1½ litre class, the first three places being taken by HRGs, but this must have been a rather hollow victory for them as the best car was some eight miles behind the Javelin. Needless to say, Jowett were delighted with this result, which had been achieved on a shoestring budget. The car had proved itself twice, and was now an international success. Everybody wanted one!

Advertising for the Javelin was original and witty, extolling the car's speed and sleek lines. There were at least a couple of dozen different ads and a typical example was:

Gay Deceiver – You've heard incredible stories about this car – stories of International race triumphs; unbelievably high average speeds. And frankly you're doubtful.

Now as you inspect her close up, you still think it can't be. She looks so comfortable, even sedate...so harmless somehow. Can that neat, tapering bonnet house such formidable power?

Then you settle down in the deep driving seat and touch the controls...and after a while you know this Javelin's been smiling at you all the time because those cars ahead seem almost stationary; and as you glide silently up behind, you realize you're travelling fast – very fast. And you brake...

Quickly the needle slips back to 40 – yes, you were up in the 70's and the whole car was smooth and steady. You didn't even notice. The torsion bar suspension holds you gently to the corners, the road seems velvet smooth, the short neat bonnet lets you see and relax at the same time and the precision steering is just that. It's all so easy in this Javelin. Now you know it. This car, so disarmingly innocent – so spacious – has all the speed of victory in its veins.

A new production line for the Javelin had been built alongside the Bradford line, with rotating cradles to receive the fully trimmed and sprayed bodies from Briggs. They were bolted into the cradle then rotated through 180° to allow the engine, gearbox, propshaft, rear axle, master cylinders and front torsion bars to be fitted. It would then be rotated back to allow the wheels, bonnet, battery and grille to be fitted. Production continued to increase, with 5,551 in 1950 and 5,769 in 1951. These figures were much better but still well short of the 150 cars per week, which the board had hoped for. The car was therefore not profitable.

It was felt money could be saved if the gearbox was built in-house, rather than buying them in from Meadows. This should not have been a problem for Jowett, as they were building Bradford boxes, and had done so pre-war. Basically this was a total disaster, because many were substandard, seizing or engaging two gears at once. There were not enough serviceable boxes being produced to keep up with the bodies arriving from Briggs. There were hundreds of bodies stockpiled around the factory and adjoining land. Due to the fall in sales, Jowett was unable to pay Briggs, so they stopped supplying bodies. There were also mechanical problems previously such as broken crankshafts and this was only resolved with the series III engine. The gearbox problem was also resolved, but sadly by now the public regarded the car as expensive, and with mechanical problems.

The backlog of cars was cleared by the end of 1952, but sales fell dramatically to 4,060, instead of the hoped-for increase. The last 380 bodies were received from Briggs in early 1953 but no more would be received. Without Javelins to sell there was no real possibility that the company would be able to survive, but they did not give up without a fight.

A 1945 mock-up of the front grille arrangement, with a couple of Jowett's workers visible. The car still had a two-piece windscreen at this stage.

An early Javelin prototype, as featured in the *War Production Record* of 1946. The booklet said: 'The car is not yet in its final form, but these are brief specifications. The engine capacity will be 1,200cc with a high top speed and 40bhp. When the new car comes on the market you will see how much we have learned, and how far we have progressed since 1939'. When the car was put into production the engine size had increased to 1486cc with 50bhp.

The second Javelin prototype DKY 463, still with a two-piece windscreen.

Another view of DKY 463. The tail of the car would be amended in the production cars.

69

The late prototype EAK 771. This car took part in the Cavalcade of Motoring in London on 27 July 1946 in front of massive crowds, including the King and Queen. This was the car's first public appearance.

EAK 771 pictured outside the London showrooms at 48 Albemarle Street. Inside is the 1913 car and the Javelin model. The car is in almost its final form, the most noticeable difference being the bonnet motif.

The last prototype registered was EKW 303, of early 1947. Here it shows the final front-end arrangement, but strangely, it has a two-piece screen, even though it was registered some months after EAK 771.

Harry Woodhead brings an early Javelin (most likely FAK 111 – chassis number two) on trade plates to the Scarborough agents, C. Ward Ltd, to allow the garage workforce to see it. Mr Woodhead is holding his hat and the car's front door handle.

The start of the Javelin production line, with a car secured in its revolving cradle ready to receive its running gear.

The engine and drive train are now *in situ*. The Bradford line is in the background.

A worker poses with a completed Javelin at the end of the production line.

A publicity shot dating from late 1949 with Javelin GAK 335 (chassis number 5209).

Another publicity shot, this time of GAK 925 (chassis number 6125), taken in London in late 1949.

Publicity shot of October 1952 with the late style Javelin registered JAK 75 (chassis number 23112). This car was displayed on the Jowett stand at the 1952 Motor Show. The car has the new aluminium grille and 7in headlamps.

A mock-up of suggested number plate and light mountings for the 1954 model. There is a drilling in the back bumper for the late number plate box, which had been discarded. This alteration never went into production.

T.H. Wisdom and Anthony Hume, 2 litre class winners in the Spa 24-hour race of June 1949, in which they beat many much larger cars.

In the pits area during the 1949 Spa 24-hour race.

A publicity photograph of the Javelin drophead produced by Carlton Coachworks in late 1948. The body structure was weakened with the removal of the roof, so the project was abandoned. This was the only one built.

The Javelin drophead at the 1948 Earls Court motor show. It was listed for several months in *Motor* and *Autocar* as an available model, but was never put into production.

Gerald Palmer pictured in the driving seat of a Javelin, the car he created, at a club national rally in Colchester in May 1989. Sadly Gerald died on 23 June 1999 aged eighty-eight.

Above and below: Crashed Javelins were often returned to the factory for repair and examination. These two examples have both had a hard time of things!

The cut-away Javelin, which shows well the use of the internal space in the car. This was exhibited at major car shows.

The factory-prepared Javelin rally car HAK 743, with some of the experimental team. From left to right: -?-, -?-, Roy Lunn, -?-, Dick Mabbett, Harold Metcalf, Bill Poulter. This car, driven by Gordon Wilkins and A. Lamburn, won the Economy Plus challenge in July 1952 by travelling 67.868 miles on one gallon of petrol. Large distances were covered by coasting the car, the grille was blanked off to keep the engine heat in and the tyres were also at a very high pressure. The closest rival was a Morris Minor, which covered 63.692 miles.

HAK 743 had been prepared by the factory to take part in the International Rally of Great Britain in April 1952. The car was driven by Marcel Becquart to a class win in the up-to-2½-litre closed class. It is seen here at the start of the Olivers Mount stage, Scarborough.

At the finishing control, Scarborough. Becquart's ultimate success was determined by the final test on the road-racing circuit of Olivers Mount.

After the success of the works Javelins in the 1949 Monte Carlo rally, and Spa 24-hour race, Javelins were used extensively by privateers. This picture shows John Eason Gibson on the 1950 Monte Carlo rally. They were non-finishers.

Tommy Harrison, Bill Robinson and Towers Leck at the Glasgow start of the 1950 Monte Carlo Rally. After Grasse there were a series of downhill sections with nasty hairpins. There had been heavy snow, which held them back and the car nearly left the road at one stage. They arrived in Monte Carlo, but sadly they were a few minutes late to be classed as a finishing team.

Charles (Berry) Leavens, Hal O'Hara Moore and Joyce Leavens on the 1950 Monte Carlo Rally. They too were delayed with snow, and were too late into Monte Carlo to be classed as a finisher.

Joyce Leavens (centre), Bea Norman (left), and Nancy Mitchell (right) entered the 1952 Lisbon rally, as an all-lady team. They finished eleventh in class and won the 'Coupe Des Dames'. Later, Nancy Mitchell as driver and Joyce Leavens as navigator would join Ford as the works ladies team.

Bob Foster and George Holdsworth on the 1952 Monte Carlo Rally set off from Munich. They would finish the rally in thirty-seventh place.

The start of the 1952 Daily Express rally with driver Wilf Parish (?) and navigator G.H. Rushton (centre), Mr Rushton says they finished third in the saloon class after having to replace a broken throttle cable.

A nice shot of Javelin EST 622. Sadly the drivers and the event are unknown to the author but confirmation of both would be appreciated.

The 1953 Tulip rally at the start of the final ten lap-race in Zandvourt. HAK 743 was the works car of Marcel Bequart, who won his class just ahead of Nijevelt in a Dutch Javelin registered NT-43-85. After the event it was discovered that there were differences between the cars of Bequart and Nijevelt and to Bequart's horror he was disqualified. He never drove for Jowett again. Nijevelt therefore won the class – and, due to the way the points were calculated, the event as a whole !

Ron Ayres in the 1954 Monte Carlo rally. He said that the Javelin travelled round trouble-free and on time. The weather had been very mild and there was not the problem with snow that had complicated previous events. This picture was taken in Paris.

86

Seven

THE JUPITER

English Racing Automobiles Ltd (ERA) approached Jowett in 1948 with a view to building a sportscar in a joint venture with Jowett, as the Javelin impressed them. Gerald Palmer was against this idea, as he felt the Javelin should be more rigorously tested, and the mechanical problems fully resolved first. This view was also held by Callcot-Reilly, so the venture did not go forward at this stage.

A works Javelin won its class in the January 1949 Monte Carlo Rally, proving its potential as a rally car. Leslie Johnson of ERA contacted Jowett again in May 1949 regarding the sports car project, and this time he was able to persuade Jowett that it was a good idea. The agreement was that ERA would produce five rolling chassis, with a developed Javelin engine. The class win in the Spa 24-hour race in June 1949 was a further endorsement that this was a viable project.

Gerald Palmer was still not in favour of the project, but in July 1949 he was tempted back to Oxford to start new design work on Riley, MG and Wolseley models. He would say later in his autobiography, 'I had fulfilled my contract with Jowett, and produced for them a worthy successor for their pre-war range of cars, which with vigorous development and competently produced, should have had a sales life of several years before major changes were necessary. I could see no future design project in view at Jowett, and I was a designer not a development engineer, so I felt justified in pursuing this approach from Oxford. Unfortunately, after I had left, the development of the Javelin did not continue as I would have hoped.'

The chassis of the new sportscar was to be designed for ERA by the Austrian designer, Robert Eberan Von Eberhorst. He had previously worked on Auto-Union racing cars from 1933-1937. The brief was to have a rolling chassis ready to display at the Earls Court Motor Show in October 1949, which was only five months away. The chassis was constructed with large diameter cross-braced chromium-molybdenum tubing, with anti-roll bars front and rear. Steering was by rack-and-pinion, which gave excellent positive light control. Brakes were originally hydraulic at the front and mechanical at the rear. These were altered to Girling hydraulics all-round very early into production, as the originals were inadequate.

The chassis was ready on time and was displayed on the Jowett stand at the motor show. ERA had also designed a coupé body for the chassis, which was rather ugly and ungainly. It was displayed at the Albemarle Street show room but Jowett knew this was not the design they were looking for! This particular car was sold to the Metropolitan Police, and it was later used at the police driving school in Hendon, but sadly no record of its fate after that is known.

Jowett knew what they wanted – an open sports with modern styling like the Jaguar XK120. The job was given to Jowett's chief bodywork designer, Reg. Korner in November 1949. He was told two cars had to be ready for March 1950, one for road trials, the other to ship to the USA for the April exhibition – it was expected the car would sell well in the USA as MG TCs, etc., were very popular there. The cars were ready in time. It was a heroic effort by Korner and his team to produce such a beautiful car in only four months.

Jowett took the bold step of entering what was basically an untried and untested car into the 1950 Le Mans 24 hour race on 24 and 25 June. The car that was entered was chassis number 4, registered GKW 111 and driven by Tommy Wise and Tom Wisdom. The car in fact won the 1½ litre class at a record speed of an average 75.8mph over 1,819 miles. The Jowett slogan for the Jupiter, 'the car that leaped to fame', was born! A class win in the January 1951 Monte Carlo rally followed.

For the Le Mans in 1951 a team of three Jupiters were entered comprising two standard bodied cars and one special light-bodied car with motorcycle type mudguards called the 'R1' Jupiter. This car was driven by Wise and Wisdom again and was very fast, reaching 100mph down the Mulsanne straight and averaging over 80mph. Sadly the car was to retire due to a blown head gasket. One of the standard Jupiters driven by Hadley and Goodacre also retired. This left the sole survivor HAK 365 driven by Marcel Becquart and Gordon Wilkins to win the 1½ litre class again! In September 1951 the 'R1' won the Watkins Glen race in the USA.

In 1952 a team of three 'R1's were entered for Le Mans. They were racing against Porsche, Gordini, Osca and Simca. These models all proved to be faster than the Jowetts, but none were to finish. The first R1 to retire, after 7hrs 18min, driven by Gatsonides and Nijevelt, failed with a broken crankshaft. The second 'R1' of Wise and Hadley retired after 16 hours also with a broken crank. This left the 'R1' of Marcel Becquart and Gordon Wilkins the sole Jowett some twenty-one laps behind the Porsche. With 4½ hrs left the Porsche was disqualified for not switching its engine off while in the pits. This then left the Jowett the sole survivor in the 1½ litre class, so it was slowly coaxed home to a third class win! Jowett realised that they were becoming out-classed by Porsche and others so did not enter the race again.

The three R1s were stored at the Jowett factory until the end of 1953 but were then stripped and dumped on the factory scrapheap with the chassis frames sawn through. An employee, Eric Price, paid £30 for enough parts to reassemble a complete car except for an engine, which he could not obtain. The car he built was fitted with a Vauxhall engine and was registered YKU 761 in 1962. At a later date a Javelin engine was fitted. Subsequently it was found that the chassis was the Le Mans winner of 1952 and its original registration number, HKW 49 was reallocated to the car by the motor taxation office. It has also since been proved that parts from all three 1952 Le Mans cars are in this car. I consider Eric bought himself a real bargain considering each car cost Jowett an estimated £5,000 to produce! The car still exists and is undergoing a full restoration.

The standard Jupiter model, known as the SA, or Mark 1, was available from November 1950, and remained basically unchanged until October 1952, when the SC, or Mark 1a was introduced. The main differences are that this later model had an opening boot, and a more streamlined hood line. Access to the boot space in the SA is only possible by tipping the seat back forward. The boot area in the SC is a little larger than in the SA but this is at the expense of the petrol tank, which is reduced in size from eleven to eight gallons. The factory records confirm 731 SA models were built, with ninety-four SC examples. There were also around sixty-nine rolling chassis produced, most of which were supplied to coachbuilders, such as Abbott, Coachcraft, Farina, Ghia, Mead, Rawson and Radford. These were built into special-bodied cars in open or saloon form and a very high proportion of these cars still survive today.

A new, cheaper, lighter model was to be the 'R4' Jupiter, to be produced in-house with Italian influenced body styling. This did not advance into production.

the 1½ litre *JUPITER*

The first Jupiter with Jowett's Chairman and the design team. Left to right: Mr. Woodhead, Messrs. Grandfield, Grimley and Korner.

As a result of the Javelin's successes, it was decided to build a race-bred, high performance car on the basis of the Javelin and so the Jupiter was born.

The Javelin's horizontally opposed four-cylinder engine was increased from 50 to 60 b.h.p. and fitted with two special carburettors; a special type rack and pinion steering gives precision steering at high speeds; an absolutely rigid tubular steel chassis gives lightness and tremendous strength; the shock absorbers have been strengthened; bigger brakes fitted and a special oil cooling system installed.

As to the body – the Javelin 5/6 seater gives way to the Jupiter's aerodynamic lightweight aluminium body – just as comfortable and as beautifully finished.

The Jupiter will accelerate to 60 m.p.h. in approximately 15 seconds, reach over 90 and cruise at 75 as a matter of course – handling the whole time in a way to delight the man who understands high-speed motoring.

At the same time the Jupiter gives complete comfort in all weathers – winding glass windows exclude draughts and the folding hood is sturdy and easy to operate.

Page 12

Details of the Jupiter as described in the booklet *50 Years of Progress*, given to all employees in 1951.

Jupiter number 1 with Horace Grimley at the wheel. This car was shown at the British Motor and Engineering Show in New York in April 1950.

This was the design put forward by ERA for the Jupiter, but it was not acceptable to Jowett, as they wanted an open sportscar. The design work for the car was given to their chief designer, Reg. Korner, with stunning results. This car was used by the Metropolitan Police and later the Police driving school in Hendon. Its subsequent fate is unknown.

LHD Jupiters, destined for the USA, on the production line in March 1952.

Hand-finishing Jupiters. The body panels were made as a set for each car, sprayed and then fitted to the body frame.

Another view of the Jupiter production line.

A publicity shot of the Jupiter showing it's flowing lines.

Jupiter chassis number 2 was the exhibition chassis used in Brussels and Geneva in 1951, as well as London in 1951 and 1952 and possibly 1953. The chassis was used to build car number 1033, the last car to be built by Jowett, which left the factory in November 1954.

Chassis number 5 was the prototype test car and was used to tour Britain and France early in 1950, driven by Horace Grimley and Charles Grandfield. It was exported to Canada in July 1950, where it still resides.

Tommy Wise at the wheel of the record-breaking Jupiter of the 1950 Le Mans. His team mate was Tom Wisdom.

Tom Wisdom (left), and Tommy Wise (right) with the Jupiter at the 1950 Le Mans. Horace Grimley is to the left of Wisdom and Roy Lunn is on the extreme left, only partially in view.

Horace Grimley hard at work preparing the car.

The pits area in Le Mans, 1950. Horace Grimley is in front of the car while Charles Granfield, in the white shirt, talks to a race official.

Wisdom and Wise were known as 'The Sagacious Two'. This was painted on the front wing of the car. Here Tom Wisdom is at the wheel.

Tommy Wise, the other half of the Sagacious Two, at the 1950 Le Mans.

For the 1951 Le Mans, Jowett produced a special lightweight Jupiter known as the 'R1' Jupiter, for Wisdom and Wise. Sadly it would retire after five hours with a blown head gasket.

Prior to retirement the car had averaged over 80mph. The class would be won by the standard-bodied car of Marcel Bequart and Gordon Wilkins.

Abberville, en route to Le Mans 1951. Horace Grimley is in the car with Tom Laycock (left) and Dick Mabbett (right).

Mulsanne straight, Le Mans, 1951, unloading the R1. From left to right: Tom Laycock, Bert Hadley (back to camera), Horace Grimley in the car, Dick Mabbett, -?-.

A pit stop in the 1952 Le Mans for the R1 of Marcel Bequart and Gordon Wilkins.

After the class win in Le Mans 1952 with Marcel Bequart driving, Gordon Wilkins as passenger. This completed a hat-trick of class wins. Jowett would not race at Le Mans again.

Tommy Wise in the RAC TT, Dundrod, 1951, where he took a class second. Stirling Moss in a C-type Jaguar is coming up close behind.

The Jupiter team ready for the 1951 Monte Carlo rally, From left to right: Bob Ellison and Bill Robinson, Gordon Wilkins and Raymond Baxter, Horace Grimley and Tommy Wise. The Ellison/Robinson car would take the class win.

Bob Ellison and Bill Robinson proudly hold the winner's trophies after their class win in the 1951 Monte Carlo rally.

The Famagusta rally, Cyprus, May 1954. Adamos Prastitis is driving. He held the Jowett agency in Cyprus, taking ninety Javelins. This car was lost under the Turkish occupation.

The author's car at the Goodwood Sprint, October 1954, being driven by K.L.W. Cook, who I assume was the owner of the car at that time. If any reader knows anything about him, I would love to hear from them!

Above and below: Publicity shots of the Mark 1A (or SC) Jupiter, which was introduced in late 1952. The main external differences are the opening boot and the more sloping hood line. The strakes on the front wings did not appear on the production cars.

The Jupiter chassis was very popular with coach-builders of the day, and around fifty were sold as rolling chassis to be bodied by numerous coach-builders. Gerald Lascelles (the Queen's cousin) bought this Radford-bodied car, and rallied it. He also used it on his honeymoon to tour the Dolomites. Another car was built by Radford but was very different in design.

Four chassis were bodied by Farina, this one for Marcel Becquart. He used it in the 1952 and 1953 Monte Carlo rallies and took a class second in the 1952 event.

Lionel Rawson bodied this car for Sir Hugh Bell and it is seen here at a club rally in Harrogate in 1967 looking a little tatty. The car still exists and awaits restoration. It was one of four cars built by Rawson.

Adams and Robinson Ltd of Sundbury-on-Thames built this Jupiter for Philip Fotheringham-Parker, a racing and rally driver. It was based on the styling of the Ferrari 212. A similar car was also built on a Morgan chassis and both cars still exist. This photograph was taken at the club's National Rally at York in 1974.

Four saloons were built by J.E. Farr and Son, Blackburn. This one was for Robert Ellison for use in the 1952 Monte Carlo rally. Ellison in bad snow crashed the car at St Flour, 130 miles from Nevers, and slid thirty feet down a ravine. The following day the car was dragged back onto the road by a team of oxen. Amazingly, the car finished the rally, arriving in Monte Carlo only a day late!

Bendall and Sons Ltd, of Carlisle, produced this car, which for a couple of years was owned by Peter Ustinov, who took it to Hollywood with him. The car is now back in the UK and is pictured here in Harrogate in 1967.

Coachcraft, of Egham, Surrey, built two saloons and possibly an open car. This example (chassis number 48) is seen at a club rally in Harrogate in 1967. The car still exists.

This open Jupiter, bodied by Gebruder Beutler of Thun, Switzerland, is regarded by many as the most attractive of all the special bodied cars. The car still exists in Switzerland.

This saloon was bodied by Maurice Gomm and was first registered in 1960. Prior to this the chassis had been stored. This is a well-known car, seen at many club events.

Eight

THE FACTORY IN PHOTOGRAPHS

The Jowett factory frontage in 1951. Note the overhead tram wires. The factory was on trolley bus wires and a trolley bus route from Bradford and many workers used them to come to work.

The main entrance to the factory. Note the winged balanced power motif above the door and the brass 'Jowett Cars Limited' registered office plaque on the steps up to the door way.

An aerial view of the factory complex. The frontage shown on p.109 can just be seen at the bottom right of the picture facing the road.

Bradfords outside the storage depot.

Javelins outside the C.K.D. compound. (Complete Knockdown – for export).

The degreasing and cleaning plant.

The machine shop.

The machine shop again. Note the crankcases to the left.

The machine shop.

The C.K.D. department, with a crated car ready for export.

Nine

WHAT MIGHT HAVE BEEN

From as early as March 1949 Jowett realised that the Bradford van needed updating or replacing as it was only a stop-gap model prior to the launch of the Javelin, and was pre-war in every aspect.

The design work for the new Bradford was given to Briggs, as the Javelin had proved expensive to build. It was felt Briggs would be able to produce a less complicated and cheaper design, with mass production in mind. The new Bradford was to be known as the 'CD' Bradford, and was to be the replacement of the 'CC' model, which had been introduced, in late 1949.

In March 1950 Charles Grandfield carried out a policy review, and was worried at what he saw. In his view the Javelin was too noisy and unreliable, the Bradford was too old-fashioned and with too small a payload area, and the Jupiter was totally untested. His proposals for the future were set out as follows:

1. Modify the Javelin in two stages. Stage 1 for the October 1951 Motor Show would be minor body changes with engine improvements to increase reliability and reduce noise. Stage 2 would be ready for the following year's Motor Show and would include a new flat four engine of stroke and bore 90mm x 90mm giving $2\frac{1}{2}$ litres.

2. Design changes to the Jupiter on an ongoing basis, as and when more was known about the car.

3. The CD commercial to be developed in two stages with stage 1 for the October Motor Show of 1951. Stage 2, known as the 'CE' commercial, would be ready for the 1952 Show and would use the $2\frac{1}{2}$-litre Javelin engine.

As the 'CD' project developed it was proposed that a full range of models were produced – van, estate, pick-up and car – and all pressings would only differ from behind the front door pillar. It was felt that a cheap car would be a success, as more first-time car owners took to the road each year. If it had been a success, there was even talk of dropping the Javelin, without a replacement. By the end of 1950 Jowett produced the first 'CD' prototype known as 'the tram' in its experimental department and it was tested extensively over the next twelve months.

Briggs were tooling up production lines for the new range and were pushing Jowett for firm orders. Jowett in turn were having problems with Javelin unreliability such as broken crankshafts, and the 'in-house' gearbox fiasco and were not able to make such commitments. During 1951 tooling costs kept increasing. In December the first Briggs prototype van was delivered – registered HKW 272 – six months late! In May 1952 an order for 2,500 vans, 1,250 pick-ups and 1,250 station wagons was finally placed. The following month, the rather stylish pick-up prototype was received and registered HKY 566. The only car produced by Briggs was received in July 1953 and registered JKU 399.

The sad fact was that mechanical problems with the Javelin, increased tooling costs, delays, and lack of finance resulted in the CD range never going into production. Only one car, one pick-up, and around eight van/estates were built in prototype form. Three estates survive in New Zealand, and one in this country. Sadly the car and pick-up were both scrapped.

Bodies for the Javelin and Bradford were no longer being received from Briggs but the Jupiter was still being produced, as there was still demand for the car. The mechanicals and body panels were produced in house so were not dependent on Briggs. The idea was formed that a new Jupiter could be produced in house at less expense than the existing model. The new lightweight 'R4' Jupiter had already been designed mainly by Roy Lunn, who was Gerald Palmer's successor. The body was to be built out of the, then experimental, laminated plastic on a new much lighter and stronger box-section chassis. It would still be powered with the Jupiter engine and gearbox, but with the addition of an overdrive on third and fourth gears. The first car built was all steel and was registered in August 1953 as JKW 537. Two laminated plastic cars were also built. The cars were very fast and had a maximum speed of over 100mph. One R4 was exhibited at the 1953 Motor Show.

It was fully tested in *The Motor* in their show issue of 21 October 1953. *The Motor* said 'The exibits on the Jowett stand at Earls Court will comprise the Javelin, the Jupiter and an entirely new sports model called the R4 Jupiter. Owing to difficulties which have arisen in the supply of pressed steel body shells, production of the Javelin is temporarily in suspense, but hope has not been abandoned that delivery will be recommenced in the not too distant future.' Sadly the supply of bodies did not commence again and the R4 was not ready for production in time, so was not to be the company's saviour. Roy Lunn was headhunted by Ford on 1 January 1954, and at the same time Jowett were in negotiations with International Harvester to purchase the business. This they did, keeping on most of the Jowett work force and utilising the production lines for tractor production.

A few Jupiters were built at the plant after the sale from unused parts, the last one being completed in November 1954. The first all-steel R4 was sold to Alf Thomas who raced it regularly at events such as Goodwood and Silverstone. The car was more than a match for the MGAs and TVRs, proving the car would have been a success had it gone into production. Thomas was able to buy the two glass-fibre cars from Jowett in 1958 and he fielded a team of three R4s for the August 1958 Silverstone six-hour race. Sadly two of the cars crashed and were damaged, so they were not raced again. Two R4s survive today, one is on the road and seen regularly, the second is finishing a complete restoration.

This however was not quite the end of the Jowett story. A promise had been made to supply spares and service for the post war cars for ten years after the end of production. Premises were found at Howden Clough, Birdstall, near Batley (to the south east of Bradford), which was a long-disused woollen mill. Albert Clegg, who had been with Jowetts since the 1930s, was the expert appointed to advise owners who had mechanical or technical problems. Spares and handbooks were available, it was also possible to have engines and gearboxes reconditioned. The company was still known as Jowett Cars Ltd., but on 18 August 1958 the name was changed to Jowett Engineering Ltd.

Blackburn Aircraft bought what was left of Jowett Cars Ltd in 1955. After Jowett's financial affairs were wound up there were no debts, and the shareholders were paid out pound for pound. Some years ago the company name 'Jowett Cars Ltd' was bought from Blackburn Aircraft, and is now owned by the Jowett Car Club, but there no plans to go back into car manufacturing at the present time!

Jowett Engineering Ltd continued to supply spares and carry out repair work until its closure on 31 December 1963 in line with the original agreement. At this time many spares were dumped, but large amounts were exported to New Zealand. Many owners then had to turn to scrap yards for spares, or track down other Jowett owners for help. Many cars were scrapped during this period as, after the closure of Jowett Engineering Ltd, there was no longer the back up for spares and servicing. Cars were worth very little, so many people found them no longer economic to run.

Having said this, the extra decade of cover offered to owners is the reason why such a relatively large number of Jowetts survive today .

What If…

What if…the Jowett brothers had not sold their shares in Jowett and had remained in control of the company after the war? It has already been noted that the Bradford was the only post-war financial success. If the company had continued to build similar cars after the war and not expanded as quickly, would it have lasted longer?

What if…the Jupiter had not been built in 1950, but more time and money had been put into the Javelin, as Gerald Palmer wanted? Would the company's finances have been better, thus removing the need to build the in-house gearbox, with its catastrophic results?

What if…the 'CD' range had gone into production in 1951, with the new $2\frac{1}{2}$-litre engine to follow in 1952? Would they have been a success, and would the Javelin have been dropped from the range. Would the proposed up-market, six-cylinder 'Venus' saloon have gone into production?

What if…the 'R4' Jupiter had gone into production in 1953 as the company's last-ditch in-house saviour? Could the company have stayed in production on a reduced basis, like Morgan?

Who knows what might have happened; one thing we do know, however, is that Jowett will be remembered as a highly individualistic motor manufacturer who did things their own way. They did not suffer the same fate as the likes of Riley and Wolseley – becoming a badge-engineered marque – but left car manufacturing with their heads held high and uncompromised.

This was the only prototype saloon produced, registered JKU 399, chassis number CD4. The car was used by Jowett Engineering at Howden Clough until its closure in 1963. It was said to have been abandoned on the derelict site, vandalized and then scrapped.

It is often asked why the CDs did not receive a better fate at the closure of Howden Clough. One reason put forward was purchase tax. Each CD in fact cost a small fortune to produce, and a would-be buyer would have had to pay the unpaid tax. With hindsight there must have been a better way to have handled the situation.

Above and below: The CD Pick-up. As with the CD saloon, this was the only prototype produced. It was registered HKY 566, chassis number CD3. It too was used by Jowett Engineering for the ten years to the end of 1963.

Some body panels were left in the open for twenty years or so, but the chassis was lost. The remains were saved in the early 1990s, but it is unlikely to be restored as there is not enough of it left.

At least four CD van/estates were exported to New Zealand. Three still survive. The fourth was rescued, but the bodywork was too far-gone, so it was restored with a Jupiter front end, and four-seater open bodywork.

Above and below: One utility, chassis number CD10, registered JKU 945, survives in the United Kingdom and is undergoing a full restoration at the present time.

The interior of the CD was rather Spartan. It had a floor-mounted gear change.

The R4 Jupiter. This model – as well as the 'CD' range – should have been Jowett's saviour, had it gone into production. This publicity shot looks like it was taken with the American market in mind. The car is complete with white-wall tyres.

Another publicity shot of the R4.

The rear of the R4. The petrol filler cap is located inside the boot.

The R4 would have been competitive, with a top speed of over 100mph, and was well capable of seeing off MGAs when raced in the mid- to late 1950s by Alf Thomas.

Nine

THE JOWETT CAR CLUB

As mentioned in the text already, the Jowett Car Club has long been recognized as the oldest one make car club in the world, and can trace its roots back to 1923 and before. From the very outset of car production in 1910 there has been a real feeling of 'camaraderie' between Jowett owners. Small groups of Jowett owners met regularly for social events and picnics, etc.

In May 1922 the *Autocar* carried an advert which read, 'owners of Jowett cars in the neighbourhood of Bradford are asked to meet at Manningham Park Gates (main entrance) tomorrow, Saturday May 13 from 2:15-2:30pm. It is proposed to have a short run to Boroughbridge and after tea, to hold a meeting to inaugurate a new club restricted to Jowett cars.' This is how the Jowett Light Car and Social Club was formed.

The following year, E.A. Dudley Ward wrote to the *Light Car and Cycle Car* magazine (10 August 1923) regarding the formation of a Jowett Club in London. He wrote, 'There are an ever increasing number of Jowett Cars on the roads in and around London, and it has occurred to me that a London Jowett Car Club might easily be formed. I have been interested to read from time to time of the doings of the long established Bradford club. Such a club must provide means of pleasant social gatherings and also opportunities for mutual assistance in obtaining the best out of one's mount.' He goes on to ask readers to contact him, if they think the formation of a club would be a good idea. The magazine carried letters from J.A. Spencer in the following two weeks, as he had already planned a London club. He published details saying that the inaugural meeting of the club would take place at Titsey Hill, Surrey on the road from Croyden to Limpsfield. The 31 August issue of *The Light Car and Cyclecar* gave a full page account of the 'successful inaugural rally of London owners, known as the Southern Jowett Light Car Club.' They stated that the club already had a brisk membership of thirty-five, where membership subscriptions to 31 December were paid on the spot!

Jowett Cars Ltd. took out an advert in *The Motor* on 18 December 1923, with the heading 'That's The Spirit!' where they gave details of the two clubs, for the north and south of the country. They went on to say: 'Would you not like to become one of this happy band of super-satisfied owners? Let us tell you all about the world's most economical car, then decide for yourself whether it is not a wonderful proposition!'

Other clubs grew up during the 1920s. In fact, the car's handbooks stated: 'Enthusiasm has been the means of forming no less than three Jowett car social clubs, and a host of little clubs or bands of Jowett car owners in various parts of the country.' These continued to flourish during the 1920s and 1930s. During the war years of 1939-1945, all these clubs ceased with the exception of the Southern Jowett Car Club, who managed to keep going with a much-reduced membership.

During the 1950s and early 1960s Jowetts were regularly seen on the road, so the membership of the Southern Jowett Car Club continued to grow from a few dozen to over 600. First a Midlands section appeared, followed by Northern, then Scotland, as members were joining throughout the country. At the Annual General Meeting of the club in 1964, it was agreed that the word 'Southern' should be dropped from the club's name to become the Jowett Car Club.

In 1966 the club held it's First National Rally in Bradford, where Jowett owners met for the day to discuss and look at each other's cars. They continued to be held on the West Stray, Harrogate for several years. In 1973 the club changed its policy, and different sections took it in turns to arrange the rally, so it is now at a different venue each year.

In the early 1960s the club was very much a Javelin and Jupiter owners club, as very few Bradfords and pre-war's were on the road. This was understandable, as Javelins in particular were still the only form of transport for many owners. Sadly the 1960s and early 1970s were a bad time for the Jowetts, as they were worth very little, so many cars were scrapped when they were no longer economical to use on a daily basis. During the mid- to late 1970s there was an increase in interest in older cars and the 'classic car movement'. More owners bought cars to restore or to use on the rally scene, while others raced them in historic racing events.

Club membership once again continued to grow, and in 1980 the club took the major step in establishing it's own spares company. They supply and also remanufacture the parts needed to keep the cars on the road. There has been a steady increase in the number of restored Bradfords and pre-war cars, and now almost 50% of vehicles at the club's National Rally are pre-war and Bradford models.

The club now has nine regional sections around the country, each with its own committee, arranging local events for members to attend. The total membership now stands at almost 600, which includes a strong overseas contingent. There are members in Australia, New Zealand, Canada, USA, South America, South Africa, Europe and even Mauritius! The club's magazine, called *The Jowetteer*, is produced for the membership each month, with technical articles, section events, advertisements, editorial comments, spares information, historical items and details of articles in the motoring press.

The club rallies now are three-day events held over the end of the May Bank Holiday, and include a scenic run, dinner dance, and concours event. There are also affiliated clubs round the world, based in Australia, New Zealand, America and Denmark, and also the Jupiter Owners Auto Club (founded in 1962) which caters for the owners of Jupiters.

Noel Stokoe is the Press Officer and Librarian of the Jowett Car Club. He would like to hear from any reader who may have owned a Jowett in the past. He would also like to borrow any old photos of Jowetts to copy for the club archive. He can be contacted at 16 Eskdaleside, Sleights, Whitby, North Yorkshire YO22 5EP.

Noel Stokoe with his 1952 Javelin and 1952 Jupiter.